Table of Contents

Foreword..2
Chapter 1, The Inevitable Energy Debacle........................4
Chapter 2, Modernization of Energy Dependence. 6
Chapter 3, Our Brief History with Petroleum....................14
Chapter 4, Signs of Energy Decay..................................16
Chapter 5, Acceptance of a Change................................20
Chapter 6, The Peroxide Engine....................................22
Chapter 7, The Peroxide Question..................................25
Chapter 8, Lunch with Jim...28
Chapter 9, The After Chapter..31

Foreword

Nothing remains the same.

Despite the instinctive routine of attaining and keeping what a person wants, everyone learns that the object of their wants will change and that is unavoidable. For the better or for the worse, it will change. Certainly, the perspective of the want will change.

The only constant is change.

With most everything in a constant state of change, people have learned to become flexible. Known as "peer pressure", it is acceptable to change as fast, if not faster, than others. The ability to welcome change and be very flexible is not common as shown by the slow timeline of drastic changes. Only when the change involves better convenience, affordability, and simplicity, is there quicker assimilation.

Don't get used to what you like.

In the context of energy, specifically nonrenewable fuels, there has been no foresight given. The change from whale oil to petrofuels occurred primarily due to the near extinction of whales coincidental to the time crude oil was being distilled nearly three centuries ago

Disjointed campaigns have been presented from time to time with much research to pacify anxious consumers with promises of plentiful fuel, yet no tangible product has resulted. Earnest promises have been forgotten.

In the past two millennia of the Modern Christian Era (0-2000 A.D.), horses and whales were the two sources of energy available for the majority of that time. It wasn't until the Steam Engine, invented near the beginning of the 18th Century, that such devices as valves and throttles became familiar. The Steam Engine was the sole and prime method for locomotion for nearly three Centuries (1705-2000 A.D.).

Diesel fuel (Kerosene) became the fuel for illumination about that time but did not fuel the Diesel Engine until it was invented near the start of the 19th Century. Refined Kerosene now fuels commercial aircraft, supertankers, nearly all tractors and trucks; its future market appears to be sound for another Century.

Gasoline requires more distillation of crude oil which wasn't done until near the beginning of the 20th Century. Before then, gasoline was considered a waste product and was disposed of as a pollutant. By implementing a Spark Ignition (SI) system, the Gas Engine was invented, and Gasoline became a fuel. Around the start of the 21st Century, 7,300,000 gallons of Gasoline are sold daily in the United States.

The lack of foresight in the development of vehicles and fuels over the past 300 years combined with the absence of planning or any organizing of future conveyances has led us to this - The Peroxide Question. The Peroxide Question will be misunderstood and entertaining before scarcity of petrofuels demands that it be asked.

Chapter 1
The Inevitable Energy Debacle

The Inevitable Energy Debacle is not a definite occurrence to happen at a predicted time in the future nor is it a topic of trendy conversation to solve a crisis. I could make a book of what the Inevitable Energy Debacle is not or what it will be in precise terms as I see it. However, the Inevitable Energy Debacle can be preempted as easily as it is being caused

Inevitable? Most assuredly. The Energy Debacle has not been discussed since it started around the start of the 18th Century. The use of kerosene distilled from crude oil depended on the drilling and pumping of oil wells of unknown quantity. At that time, instruments to measure the size of subterranean oilfields did not exist. Such instruments were not necessary because the oil kept pumping year after year, besides the oil market was not fully realized. Had the oil been pumped at a sustainable rate of flow, the energy debacle might not have been inevitable.

Energy? This word has been overused to the point of vagueness. For purposes here, energy is in the form of a liquid fuel that resembles gasoline or kerosene. Since flammability of gasoline or kerosene is well known as mpg (miles per gallon), mpg is used as a standard for comparison, however, a future fuel is not limited to the mpg of gasoline or kerosene.

Debacle? Absolutely! Not for everyone but definitely for the majority of people that depend on other entities for their needs. Entities, such as governments, are assured of plentiful energy since they control distribution, sale, and taxation. Entities, such as petroleum corporations, are assured of plentiful energy since they produce the energy at the rate to produce planned profits. Entities, such as those that depend on petrofuels, are not assured of plentiful energy after the year 2040 A.D., estimated.

There are a significant number of people that avoid debacles, cope with debacles very well, or completely ignore debacles by changing their level of dependence on what caused the debacle. Of these, ignoring the debacle is the most difficult since debacles have contingencies that impinge on other matters. It is impossible to list all the resulting debacles and contingencies of the Inevitable Energy Debacle here.

Anyone can dismiss the Inevitable Energy Debacle on the basis that it has already happened with the Oil Embargo of 1973, but I say that was handled as less than a debacle. In that situation, Arabian oil producers halted oil exports for political reasons: that is less complicated than the depletion of crude oil. Strategic Oil Reserves were opened, and the problem was solved by domestic production.

There were massive hardships caused during that Oil Embargo of 1973; principally because it was the second Oil Embargo of Modern Times and directed at the United States. The first Oil Embargo of Modern Times was the Oil Embargo of 1940 which was directed on Japan and caused, in part, the Pacific Theatre of World War Two.

Theoretically, Pearl Harbor, the Nuclear Bombs, and most of the war tragedies that happened in the Pacific Theatre of World War Two would not have happened if Japan had not had their oil supply embargoed. Although it is unlikely that a third Inevitable Energy Debacle would cause a third World War, it certainly could become a factor that no one wants.

Chapter 2
Modernization of Energy Dependence

Energy Dependence is requisite to Human survival at its present level and determining a date for when it became "modern" is dubious, at best. Compounding matters are the fact that acknowledgement of Energy Dependence is unpleasant to realize and the fact that there have been varying degrees of assumption about how much Energy is needed certainly obfuscates the issue.

Many references of the Western Civilization are based upon the Roman Empire which existed from 100 B.C. to 400 A.D. Since Rome had a system of distributed water (aqueducts) and established paved roads, the year 100 B.C. is chosen as the beginning of the era when Energy Dependence became "modern".

The formal use of Mechanical Energy made possible by the Steam Engine was still 1700 years away, but wood and coal were delivered to Rome to warm and light houses.

The first commercial Steam Engine was the Savery Steam Engine invented by Thomas Savery and patented in 1698 A.D. It operated by developing higher than ambient pressures to drive water out of a sealed container. The Savery Engine had no piston or timing mechanism and could barely manage one horsepower of force. It was sold as a pump to clear coal mineshafts in England of water and had very few other functions.

Savery met Thomas Newcomen shortly thereafter and in 1712 AD, the Savery patent protected the Newcomen Steam Engine design. The Newcomen Steam Engine had a piston and cylinder to develop mechanical motion as opposed to fluid displacement. The Newcomen Steam Engine was more powerful than the Savery Steam Engine, but it was also very inefficient with fuel. Despite its shortcomings, the Newcomen Steam Engine became the common steam engine for all of England and Western Europe for the next fifty years. It was, essentially, the only mechanical energy source other than a water wheel that was available at the start of the First Industrial Revolution.

The First Industrial Revolution began in 1760 A.D. and continued through 1840 A.D. A Second Industrial Revolution, known as "The Technology Revolution", followed and ran for 1870 AD until 1914 AD when the First World War ended it. Neither Revolution involved the public distribution of utilities or the public sales of cars: that wouldn't gain prominence until after the end of the Second World War-1945 A.D.

Although the First Industrial Revolution is generally discussed about with Great Britain in mind, France, Germany, and many other Nations participated as well. It quickly expanded to the United States with the Textile Industries when Steam Engines motorized factories and the water wheels became obsolete. Horses treading on circular treadmills or towing a pitman on a circular path also became obsolete at about that time.

During the First Industrial Revolution, James Watt improved the efficiency of the Newcomen Steam Engine design with the addition of a sliding valve and a crankshaft eccentric journal. The Newcomen Steam Engine also became obsolete quickly.

Around 1782 A.D., the Watt Steam Engine became the new standard by which all other machines were driven. Steam Boiler Operators transitioned with training to become Steam Engine Operators. Without much delay, Steam Engine Operators powered up entire textile factories by means of a driveshaft and flat belt system. In the Textiles Industry, multiple looms could be turned on and off by engaging/disengaging the clutch that controlled the flat belt.

Steam Engines replaced Water Wheels as the preferred source of Mechanical Energy. Water Wheels and Dams have been used for Mechanical Energy (Hydrodynamics) since the first ones were built between 300 B.C. and 100 B.C., but droughts and flooding have always been their weaknesses.

With a steady supply of coal to heat the boiler, the Watts Steam Engine could operate all year around. The total power output of the Watts Steam Engine had never been realized before and the possibilities for its use were enthusiastically realized.

Not long thereafter, Steam locomotives and steamships were built to replace horses and sailing vessels as a better mode of transportation. Bulk transport of nearly all commodities could be accomplished by steam propulsion on land or water. It may have been proper to assume around the end of the 19th Century that Steam Engines would provide all the Mechanical Energy ever needed, however, progress is an insatiable challenge for engineers and scientists.

Steam Engines prevailed as the prime movers of the First and Second Industrial Revolutions and coal was the principal fuel. Pumps were developed to transfer kerosene and room was made for kerosene storage tanks because the advantages of a liquid fuel are many compared to shoveling coal. Although coal remains affordable and is still used in stationary power plants, liquid fuels, both gasoline and kerosene, are the preferred fuels for both mobile and stationary engines.

Shortly before the 20th Century, Rudolf Diesel invented the first

Internal Combustion Engine-the Diesel Engine-and it operated with a distilled petroleum fuel that was previously only useful with lanterns-kerosene. The introduction of an Internal Combustion Engine into the modern foray of achievements would have unforeseeable advantages and consequences.

The Diesel Engine used "compression ignition" (CI) to cause a kerosene and air mixture to combust shortly after Top Dead Center (TDC) on the second revolution of a Four-Cycle Engine Sequence. The reciprocating piston of each cylinder pushed against the Journal on the Crankshaft to develop rotary motion. Rotary motion has always been the preferred output of all engines and motors.

The power output produced by Diesel Engines exceeded expectations, the convenience of using kerosene, and the higher efficiency of the Diesel Engines all proved to be advantageous over Steam Engines. Ships and Locomotives were fitted with Diesel Engines beginning in the Early 20th Century.

Although the use of Early Steam Engines persisted through the first half of the 20th Century, they eventually became obsolete. In addition to the noise and trouble of a working Steam Engine, there have always been the hazards of Boiler Explosions which took many lives.

Numerous "Spark Ignition" (SI) engines have been invented beginning in the late 18th Century, around 1790 A.D. SI Internal Combustion Engines resemble Steam or Diesel Engines as they are all reciprocating engines.

Early Spark Ignition Engines operated without a modern Throttle that could control engine speed. Once the engine attained the desired speed, the spark was interrupted, the engine slowed, and the spark was re-established. These engines were known as "Hit and Miss" engines, they were popular on farms in the early 20th Century, and the only alternative power available at that time was a horse.

An interesting Historical fact related to gasoline-fueled SI engines is the Cuyahoga River Fires that happened near Cleveland, Ohio. The Standard Oil Company operated petroleum distilleries there for kerosene distillation. Gasoline was a nuisance waste byproduct because SI engines were not popular for many years.

Consequently, Standard Oil dumped the surplus gasoline into the Cuyahoga River. Thirteen Cuyahoga River Fires happened before environmental action was taken in 1969, about 60 years after SI engines became popular with the introduction of the Ford Model T.

Incidentally, the Cuyahoga River Fire of 1969 is credited for inciting environmental consciousness and promoting the concept that led to the Environmental Protection Agency. No alarm was made about dumping gasoline into a river until then.

Over time and at the behest of the Standard Oil Company, automobile manufacturers elected to build cars with SI engines. With a controllable Throttle and plentiful, cheap gasoline for fuel, the American Automobile market has grown enormously from October 1, 1908 when the Ford Model T was introduced through today.

A fourth engine, the Turbine Engine, was first demonstrated as the propulsion engine of the German war aircraft, the Messerschmitt 262, a.k.a. ME-262. The Turbine Engine was the first engine design that was not "reciprocating"; meaning no crankshaft is needed to produce rotary motion. Therefore, the output of a Turbine Engine is consistent and smooth.

The Turbine of the Turbine Engine operated for years previous as the impeller of Hydroelectric Power Plants and modern Steam Engines, yet it wasn't until the Messerschmitt 262 that Turbine Engines gained attention as a superior engine that made jet aircraft possible.

Many improvements have been made to Turbine Engines over the years, yet the fuel that powers them remains the same-kerosene. Originally, the mass and speed of burnt gasses exiting the Turbine Engine acted as the thrust that powered the aircraft. Nowadays, the impeller of the Turbine Engine drives reduction gears that turn a larger fan in a High Bypass Turbofan Engine configuration.

Many engines designs have been made in the 20th Century but only three are in commercial use-CI, SI, and Turbine. Three of the newer designs are known as the Quasiturbine, the Six Stroke Engine, and the Wankel Engine.

Of these three new designs, only the Wankel Engine began commercial production as the 1970 Mazda R130 car and ended with the 2012 Mazda RX-8 car.

Other than Rocket Engines which operate on liquid Oxygen and liquid Hydrogen, Solid Fuel, or some other formulation, no engine or fuel has ever been offered that would disrupt the standard petrofuels of the past three Centuries. Electricity has made limited inroads as hybrid vehicles operated by gasoline engines charging batteries that operate DC electric motors but the plan has been haphazard

Although some new engine designs have challenged existing engines, all have been designed to operate on either gasoline or kerosene for a fuel. There has never been a fuel to contest petrofuels.

Another one of the early applications of Mechanical Energy was to generate Electrical Energy to drive Electrical Motors. Electric Motors were used, sometimes, to turn Generators which generated the Electricity that illuminated streets, factories, and houses. As the demand for Electrical Energy grew, Steam Turbines and Hydroelectric Dams were built to power the larger Generators.

As the first Man-made and controllable Energy Source, Electrical Energy was received favorably. Many conveniences were built to operate off the growing Electrical Grid; first in urban settings, then onto rural settings. Eventually, Electricity in the United States became ubiquitous.

Methods to generate electricity have become increasingly complex and now include photoelectric grids to make use of the Sun for up to twelve hours a day.

Substantially less electricity is generated on cloudy or stormy days and there is no way to stop excessive generation other than turning off sections of the entire system.

Photoelectrically-generated electricity must be Direct Current (DC) electricity of low voltage with the photocells available today. This original electric current must be conditioned into Alternating Current (AC) with a frequency of 60 Hz by means of an Inverter in order to integrate with the existing current of the National Electric Grid.

Subsequently, the Root Mean Square (RMS) factor of AC is .707 that of DC so nearly 30% of the electricity generated is lost in this initial conditioning of the electricity generated. Additional losses occur with the photocells and the distribution of current, in that, photoelectricity is not as easy and free as it is generally advertised to be.

With so much attention directed on the generation of electricity to power the National Electricity Grid, flippant disregard should be expected for the transportation fuels market-kerosene and gasoline-which has proven to be so reliable for the past Century.

Although the past 20th Century has been one half of time that kerosene has fueled all engines, including the boilers of steam engines, the last five decades of the 20th Century is approximately all of the time that gasoline has fueled all SI engines; the majority of which power light cars and trucks.

Chapter 3
Our Brief History with Petroleum

The first use of Petroleum occurred before recorded History; perhaps as far back as a million years or more. Most crude oil was buried hundreds of feet underground but there were exposed Tar Pits. It is logical to assume that early Homo Sapiens were probably curious to use them.

Coal is technically a solid state of Petroleum and it was formed 65,000,000 years ago about the same time crude oil began to form. The Theory for the formation of most hydrocarbon fuels is that the K-T Extinction Event ended the majority of living matter, both plant and animal. The remaining organic, carbonaceous matter became a sedimentary layer of the Earth's crust.

Coal and crude oil existed without purpose for millions of years until people began to integrate them into their daily lives within the past few thousand years. The mechanization of coal and crude oil extraction processes did much to improve the use of coal and crude oil in the past few hundred years.

Better mining practices, begun in the last few Centuries of the Second Millenia of the Modern Christian Era, have assured an ample supply of coal for the next millennium (3100 A.D.) is one date of many projections given to estimate the total depletion of coal. The actual date will have to be determined by future generations.

Future discoveries of coal might postpone that date (3100 A.D.), however, increased rates of consumption of coal will certainly abbreviate that projection. Increased efficiencies of use will extend that date and continued wasteful practices will certainly decrease that projection.

At the present rate of consumption, petroleum reserves are calculated to last, at least another 100 years (about 2100 A.D.). As with coal, increased consumption of crude oil will abbreviate that projection and new discoveries of crude oil will extend that projection. Increase efficiencies of use will extend that date and continued wasteful practices will decrease the projection.

With only one brief Century remaining before total depletion of crude oil is projected, no concerted effort is presently underway to avert certain disaster; it is an inevitable energy debacle! Most debacles can be ameliorated with planning
but no planning for any inevitable energy debacle is presently of public concern.

Presently, the promises of hybrid cars and other distractions alleviate any general concern over petroleum depletion. The price increases that accompany dwindled supplies of any commodity are seldom acknowledged with petroleum. The nuisances of higher gas prices are, generally, not associated with personal finances.

Chapter 4
Signs of Energy Decay

Physically, the Earth has less Energy than it originally did. All of the coal, trees, and other combustible materials used over the course of the History of Man are now ash and inert gases. Since combustion products do not accumulate in certain places, they appear to be non-detectable to any passive observer.

Eventually, a dwindling of Energy availability will have to occur. Denialism of Energy Shortages are predictable; so is conservatism of Energy to make what is present last longer.

One instance of Energy conservationism is not enough to alarm anyone into anxiety over future Energy availability. In fact, innumerable instances of dwindling Energy supplies will not alarm anyone provided they have all the Energy they want. Incipiently, people have adapted to less Energy and have learned to live with less Energy many times.

A willful diversion to the prospect of less Energy is the increased willingness that people have developed to adopt and use new technology. For example, whales were no longer slaughtered for their oil to illuminate houses and offices after petroleum distillation was commercialized and kerosene became affordable.

Whaling is an apt illustration of how one Energy Debacle was ameliorated. It is used here to show that our collective use of Energy causes unconscionable damage before a practical alternative is given effort. From nourishment to illumination, extinction of whole species of whales was an acceptable solution so long as Human desires were satisfied.

Although whaling was done over 5,000 years ago (3,000 B.C.-1950 AD), it was done for sustenance and is thought to not have had a great effect on the Whale population. Harvesting whales is obviously dangerous and sailing vessels of the Early Whaling Era certainly gave the advantage to the whale.

All of that changed in the 16th Century A.D. when a profit could be made by whaling, the ships became more substantial, and there was no other fuel for lighting other than whale oil. Sailing began to venture further away from the Mediterranean Sea and oceanic shores. The myths of sea monsters and the superstitions about sea gods devouring seamen withered.

All types of whales were harvested between 1500 A.D. and 1900 A.D. with as many as 50,000 whales killed in one year. There were no regulatory agencies involved with whaling until the year after World War Two (1946 A.D.) when The International Whaling Commission was instituted. Consequently, whales were still harvested but with respect for replenishment of whale stock. As the market for whale oil waned in competition with the newly refined kerosene, enforcement of whaling regulations became more possible.

Another sign of the dwindling supply of Energy is Hydraulic Fracturing, otherwise known as "Fracking". Hydraulic Fracturing of subterranean rock, hundreds of feet below the surface, to release methane gas for extraction was first tried shortly after World War Two but did not become a dominant method of extracting Natural Gas until 40 years later (1985A.D.)

Fracking did not become a practical method of extraction until after the nearer and more convenient pockets of natural gas began to dwindle.

The massive system of gas pipelines used to deliver Natural Gas could not be abandoned-that would be unthinkable! Therefore, Fracking became the new technology used commonly to salvage an expiring industry and source of Energy.

Fracking was first tried and found to be effective in Oklahoma and Northern Texas. The use of subterranean Nuclear bombs proved to be less efficient and less practical. With experience, improved proppants and practices has made Fracking the standard operation to access Natural Gas deposits until they once again begin to dwindle.

A third sign of the dwindling supply of Energy is the recent theory of "Abiogenic Oil", a.k.a. Adiabatic Oil, conceptualized around the first decade of the 21st Century A.D. It challenges the old theory of Biogenic Oil Theory which asserts that all oil began when organisms-both plants and animals-were killed 65,000,000 years ago during the K-T Extinction Event. The Biogenic Oil Theory stood undisputed for hundreds of years until the Abiogenic Oil Theory was proposed.

"Abiogenesis" basically means "begun without living matter" and it theorizes that Carbon and Hydrogen are present in the Earth's core, possibly deposited there by a meteorite billions of years ago during Earth's formation. The lighter mass of Carbon and Hydrogen is not to be debated.

The concept of Adiabatic Oil was welcomed. In that concept, it was proposed that Oil constantly seeps upward against gravity and is captured under a non-permeable cap of sediment until that cap is pierced by a well drilling bit. Because the future quantity of oil is not known, the future quantity of oil should not be of concern to anyone or, so it is supposed.

Through a process unique to Earth, the Carbon and Hydrogen present at the Earth's core combine to form Hydrocarbons under intense heat and pressure as they migrate upward against gravity. In the presence of additional Carbon, the light Hydrocarbons merge to form heavier hydrocarbons and so crude oil is made.

The heavy hydrocarbons travel along subterranean fissures and through faults until non-porous rock traps them to form an oilfield. Many oilfields are located five or more miles underground which portends that the oil traveled upward rather than downward and under non-permeable rock.

Silicon is suspected of accelerating the chemical synthesis of Hydrocarbons catalytically. Granite, quartz, and other silicon-bearing minerals seem to assist the Abiogenic process as the Hydrocarbons travel hundreds, if not thousands, of miles away from the Earth's core.

Despite the many facts that tend to validate the Abiogenic Oil Theory, it remains a Theory. The Biogenic Theory of fossils forming fuel that has served us well for Centuries remains intact. Abiotic oil is still a theory. The quantity of abiotic oil available is not known to be sufficient for a world population of 7 billion people for a known amount of time or not.

Chapter 5
Acceptance of a Change

The automobile and all vehicles fueled by petroleum distillates have only been of use to us since the Diesel Engine began the evolution around the start of the 20th Century as a stationary power plant. It soon was mounted to ships and submarines as the motive device, then trucks, tractors, and trains some twenty years later.

Despite Vehicles operated by Steam Engines and Electric Motors existing at the same time, kerosene as a fuel is given significance here. Kerosene was used to fuel the Boilers used to power modern Steam Engines but that is of small significance here. Kerosene was already established as the first liquid fuel because it was used in kerosene lanterns as the only source of light after the canvas torch

Progress on the Diesel Engine (CI or Compression Ignition engine) happened fast in the first half of the 20th Century. Never before in the History of Man had so much mechanical energy been possible and the opportunities it presented were quickly realized. Ships and trains became larger and carried more cargo. Airplanes with Diesel engines flew great distances without refueling. The most powerful truck (with a 160 hp Diesel engine!) was built in 1932. New designs of airplanes became possible with kerosene powering Diesel engines.

A distribution system for kerosene barely kept pace with the burgeoning use of kerosene. Airports, seaports, and truck terminals were rapidly constructed with fueling stations before pipelines were built to keep up with the demand. It wasn't until after World War Two that service stations began to become ubiquitous but they principally fueled

SI engines with gasoline.

Gasoline engines (SI or Spark Ignition) existed in Germany beginning in the last quarter of the 19th Century and they ran on coal gas. Essentially, the Gasoline Engine did not come into prominence until the Ford Model T coupe was built from October 1, 1908 to May 26, 1927. Most all car manufacturers understood the necessity of keeping with one type of fuel and therefore one type of engine-the SI engine.

Both CI and SI engines share many similarities since both are "reciprocating" engines. The rotation of either engine is accomplished by a crankshaft which is notoriously inefficient and a common cause of engine failure. Over time or during short periods of high power, it is the crankshaft that fails, particularly with CI engines. Consequently, crankshafts are built extra heavy and the injectors are limited in flow rate to restrict the rpm (Revolutions per minute) speed to prevent overrevving the engine.

Despite dramatic advances in technology, no engine or fuel has ever been tried that would compete with what we have had since beginning of the 19th Century. Indeed, there doesn't seem to be any reason to wonder about what we have now because it has worked so well for such a long time: at least five generations have venerated the current state of mechanical energy generation as it stands now.

Chapter 6
The Peroxide Engine

The Peroxide Engine does not exist and is not thought of because there is no Hydrogen Peroxide fuel or concept of using a manufacture able carbonless fuel. With licensure to buy and safely handle 50% Hydrogen Peroxide, it can be purchased at roughly twice the cost of retail gasoline-$6.00 per gallon in 2018 in a 55 gallon plastic drum. It's way too early to debate the cost benefit of gasoline over peroxide since that might change with the economics of volume besides 50% HOOH might be too concentrated and a lower concentration might suffice for combustion.

In this present quandary of exploration, only existing 4-stroke SI engines of 2 liters displacement or more are available for experiments. HOOH might combust too rapidly and destroy CI engines and Turbines Engines are too expensive. Also, Turbine Engines don't allow for the compression ratios or combustion dwell necessary in their present configuration.

An SI reciprocating engine is the least desirable, but most familiar, engine type to experiment with in determining Peroxide Concentrations and associated factors, such as range per tankful. The Peroxide fuel may detonate without the use of a SI system or with a modified SI system. The exact mechanics, performance, and feasibility are yet to be known.

Perhaps I got ahead of myself with offering a Peroxide Engine when a Peroxide Transmission is also not known. Many automobile systems that worked fine with gasoline might be affected adversely by the first fuel ever conceived that doesn't depend on natural resources. Living without Carbon Monoxide and Carbon Dioxide emissions is a factor that also might affect automobile systems.

In light of such an obviously difficult and expensive task as engineering a Peroxide Engine by using a 2.0 Liter SI Engine, a simpler and more economic beginning would be attempting to demonstrate the feasibility of using HOOH as an oxidizing agent of existing SI engines using gasoline. By using HOOH to increase Percentage of Oxygen in Combustion, gasoline might be extended and horsepower might be increased.

This much simpler approach for demonstrating HOOH as an oxidizer will start the Transition Phase required for a Peroxide Engine. As the gasoline use is lessened, the combustion of Hydrogen Peroxide will be proven.

In the Transition Phase, the 21% Oxygen Component of the Atmosphere will be increasingly supplemented to a practical percentage for better combustion. Turbochargers and Superchargers cannot accomplish this as they only densify the available oxygen percentage of air. The entropy of the Hydrogen Component of HOOH can only enhance the combustion process. The H2O component of the HOOH solution might be detrimental, at least in an SI Engine.

This process is done by injecting a Hydrogen Peroxide mist into the SI Engine Intake before the Throttle Body Injector (TBI) where it mixes with gasoline. Because HOOH and gasoline combusts on contact, a minimum rpm must be reached that assures that reaction doesn't happen before the Intake Valve. Otherwise, a "backfire" will occur with fire in the Intake Manifold!

In a car or light truck, a Peroxide tank and pump would deliver a certain percentage of Hydrogen Peroxide to the Intake Manifold and the economics of operation would determine the feasibility of the extra work and expense. A variable nozzle could increase rate of delivery of HOOH per rpm and other systems are known that would increase the efficiency of the system.

The Transition Phase is exploratory and discoveries might change the course and outcome of the Transition Phase. It is important that the first purpose of the Peroxide Question remain foremost before the existing quantity of Petroleum dwindles to the point of desperation.

Of course, there are plans and measures to prevent such a crisis from happening. It's just that it isn't of any concern at the moment (2018) and no one knows of them!

Chapter 7
The Peroxide Question

Will Hydrogen Peroxide be the Next Fuel after Oil?

There has not ever been sufficient reason to make the Peroxide Question entertaining and pleasant. Enough courage to engage such an expansive question with so many detriments would be difficult to garner without a National Tragedy, such as Pearl Harbor or 911, to force attention to an inevitable energy debacle. Interest in the Peroxide Question is unlikely without the loss of several thousand lives.

Everyone will need to phrase the Peroxide Question in such a way that it appeals and is understandable to the queried individual. To start, the Peroxide Question should be asked of yourselves before discovering how obnoxious it will sound to an unsuspecting person

An introductory phase might be- **"What will we use for fuel after kerosene and gasoline are near depletion and expensive?".** That type of phraseology is gentle and so ambiguous that few people will need feel threatened in answering it. Quickly follow that question by the Peroxide Question-**"Will Hydrogen Peroxide be the Next Fuel after Oil?".** That will alleviate the assumption that there is no substitute for petrofuels.

Few need to know that that question is forty years after the PPOE-Post Peak Oil Era-began to answer it. Nearly everyone thinks it's a new question and something to play with. Most everyone is too busy to be disturbed with a subject that must be tended to by politicians and scientist of world acclaim or, so it is supposed.

The Peroxide Question must be simple and dispel the notion that Hydrogen Peroxide is fuel strictly manufactured for rocket engines. Actually, Hydrogen Peroxide is a lesser fuel than Hydrazine and many modern solid rocket fuels.

A Peroxide Question of such intention would be directed as such- **"Since Hydrogen Peroxide hasn't been the best rocket fuel since its use in the V-2 Rocket of World War Two, don't you think it could diluted with distilled water and used to fuel cars?".** Although lengthy questions are to be avoided when trying to entice curiosity and history snippets tend to confuse the queried, it is apropos to make best use of your time when questioning.

A **Peroxide Question meant for juveniles, some of whom might have relevance to the answer, should be simple, general, and sound like this- "Have you ever heard of Hydrogen Peroxide?".**

While it is never appropriate to interrupt children of any age, some children are inquiring and might be very well informed about chemicals, so you should be prepared for a lively discussion. Also, Hydrogen Peroxide will be less obscure to future generations than any of the preceding generations.

A Peroxide Question meant to be challenging would take the form similar to- **"Do you have any idea what we will use after kerosene and gasoline become depleted?". This type of phrase is the Peroxide Question in its most direct and simplest iteration. The answer is always the same-Hydrogen Peroxide.**

Having said that, every learned and unlearned person is goaded to dispute Hydrogen Peroxide as a fuel and challenge the rationale I used to draw that conclusion; the first rationale that I used is that all previous fuels have failed against the splendid qualities inherent to the carbon fuels that have served so well for the past Century.

I can't list all the failed fuels that have been tried; I'm sure very few people know them all. Indeed! Discussion about failures is generally a waste of time. If a fuel candidate existed that could supplant petrofuels, other than Hydrogen Peroxide!, it is very plausible that no interest developed in it.

Ammonia (NH3) teases in its claims to be a fuel but it has Nitrogen, an inert gas, as its binding molecule. Coal gas was tried in the early years of engines with some success, but no fuel can compete with the distribution network of kerosene and gasoline as it stands after more than a Century of development. Perhaps the greatest obstacle will be the typical desire many people have to stay with what they trust, at any cost.

Chapter 8
Lunch with Jim

Mr. James Bede, (b. April 17, 1933-d. July 9, 2015), was a nearly lifelong hero to me. In my eyes, he was the most prolific American aircraft designer of all time having built seventeen aircraft designs, as well as, a fairly good humorist. Despite his world-famous reputation, he remained humble and always made time to visit with me at the EAA Convention in Oshkosh, Wisconsin.

The last time I visited with him there must have been when the BD-17 rolled out and our talk was about aluminum honeycomb panels. Mr. Bede showed that flat panels could fly years earlier with the BD-4 but the aesthetics of compound curves and glossy paint were necessary for aircraft manufacturers to sell expensive products. Mr. Bede managed to make a living by designing airplanes that flew safely without the frivolous expense of difficult construction with exotic materials.

When I heard that Mr. Bede moved to Medina, Ohio, I called him and asked to meet with him. I was surprised that he remembered me, and he was quick to extend an invitation! It was going to be wonderful to be able to talk one on one without hundreds of aircraft homebuilders interrupting. We agreed to a time near Christmas 2012.

When I got there, I was welcomed in the typical Bede manner and offered a seat next to his Computer Assisted Drafting (CAD) station. I had to watch my step because his office was cluttered with all types of aircraft parts, aircraft models, and boxes and boxes of papers waiting to be filed in several cabinets.

Mr. Bede had recently undergone hip surgery and moved slowly with obvious pain, but he tried as best he could to not let it interfere with our meeting. He gave me a label pin that looked like a BD-5 and offered me a coffee before asking me why I made the trip.

In the presence of my Hero, I blathered about how his work spurred my imagination as a pilot, especially his BD-5 design. I told him of my drawings of a similar design that I had that took up so much of my time in Grade School. He seemed to enjoy listening to my story before showing me his newest design on his CAD station. It was a Flying Car- the BD-18-like none other! The hours slipped by and it was time for lunch.

I drove us to a restaurant he liked well enough to climb up the front steps with his sore hip and he picked a table that overlooked an idled manufacturing plant across the street. He told me the stories of the restaurant and the plant and when done with those, he began the story of the 1962 Chrysler Turbine Car.

Besides his gregarious nature and encyclopedic memory, Mr. Bede pounced on most any subject once he found another person with a similar level of interest. His knowledge of the Turbine car took us through lunch and desert until the restaurant was nearly empty again.

I noticed he looked a little tired when I dropped him off at his office after lunch. His hip was plainly aggravating him. He still managed to answer a few questions about the business prospects of his Flying Car and that stirred him back up to his typical ebullience. I left with a good hope of being able to buy a Flying Car in my lifetime.

On the drive home, I thought of what Mr. Bede had said about why the Chrysler Turbine Car never went into production. My assumptions about costs and precision assembly were quickly determined to be wrong by him. All my guesses were wrong.

Mr. Bede assured me that it was the Mechanic Problem, or rather the lack of Turbine Mechanics Problem, that prevented the Chrysler Turbine Car from reaching production. All of the benefits of a turbine engine outweighed the consequences but the foresight of training mechanics to repair a turbine engine was too daunting.

Also, most turbine repairs involve expensive instruments to assure the mass of the turbine is perfectly balanced. An unbalanced turbine can violently explode and send shards of metal flying in all directions. Sensors can be built to prevent this from happening, but people find ways to override or ignore sensors. The training and risks of any engine, besides what is known about reciprocating engines, was just too great of a dilemma to bother with while a satisfactory engine served the same purpose.

Mr. Bede taught me that excuses, real or imagined, halt many fantastic machines from being built. An excuse for running out of fuel to power those fantastic machines won't be necessary until after there is no fuel remaining and then it will be too late.

The reciprocal issue to that is there might not be enough brave people to support a new market of fantastic machines, possibly dangerous machines like the Ford Model T. The urbanization of people enhanced the appeal of the Ford Model T which, in turn, accelerated the obsolescence of the horse. An enigma occurs for what would have become of automobiles, if mechanics hadn't learned that trade.

Chapter 9
The After Chapter

I understood twenty years ago that books about dismal subjects are not entertaining or bestsellers. That is sufficient reason for procrastination and I can find many more reasons, but millions of people are ignoring a debacle better than me.

Producing 7,300,000 gallons of Hydrogen Peroxide daily is possible if that's what it takes to keep everything as we know it today. Horses and bicycles will still be available for those who can use them in the future and I expect several thousand people will not hesitate to pay more than $10 for a gallon of gas. $20?

A better engine with a better fuel is possible now; has been since the start of the 21st Century yet the newest global fascination is traveling to Mars, ending pollution and waste, or politics and international intrigue. There isn't any national crusade enveloping our desires as there was in the 1960's with NASA.

The most recent Federal Agency is the Department of Homeland Security which envelopes our fears of dangerous terrorists and, at best, can maintain what we have in its present condition. Technology cannot be developed when threatened. Progress can never be defensive.

The old adage- "Don't fix what's not broken."-has had its place in our History and it serves a general propensity of completely emptying a commodity, such as petrofuels, before adapting to living without it. Imagine leaving whole reservoirs of crude oil for future generations; that isn't preposterous!